T0026985

THE INSIDER'S GUIDE TO

ENGINEERING

Jess Dixon, of Andalusia, AL, got tired of being tied up in traffic jams, so he designed and built this novel flying vehicle. It is a combination of automobile, helicopter, autogiro, and motorcycle. It has two large lifting roto[r]s in a single head, revolving in opposite directions. It is powered by a 40 HP [engine] which is air cooled. He claims the machine is capable of speeds up to 100 miles an hour.

-*Modern Mechanix,* Nov. 1941

THE
INSIDER'S
GUIDE TO

ENG
INE
ERI
NG

LUKE ZOLLINGER
P.E.

Copyright © 2023 by Luke Zollinger

Edited by Andrew Durkin
Cover and interior design by Masha Shubin | Inkwater.com
Cover and interior design revisions by Tessa Schmitt

All rights reserved. No part of this book may be reproduced or trans-
mitted in any form or by any means whatsoever, including photocopying,
recording or by any information storage and retrieval system, without
written permission from the publisher and/or author.

Paperback   ISBN-13 978-1-66640-355-8
eBook   ISBN-13 978-1-66640-356-5

3 5 7 9 10 8 6 4 2

CONTENTS

ACKNOWLEDGEMENTS

A special thank-you to all who contributed to the development of this book:

Kevin Beauregard
Roberto Albertani, Ph.D.
Tom Ohnstad
Cameron Crawford, Brigadier General, USA (Ret.)
Todd Fronek, Larkin-Hoffman Attorneys
Mark Miller, P.E.
John Parmigiani, Ph.D., P.E.
Jay Cobb
Patrick DiJusto
Sahara Peterson
Andrew Durkin
Masha Shubin
Paul Cheney
Vernelle Judy
Tim Harris
Tom Doherty
Ginger Bock
Tessa Schmitt

I F YOU ARE AN ENGINEER, ENGINEERING STUDENT, OR A
regular person with engineering curiosity, this is
the book for you! Written for all audiences interested in engineering, this book will inform
or refresh you on the basics of the profession
and clue you in to interesting facts – things
you don't know or perhaps forgot. There is
something here for everyone and you will
enjoy getting a better idea of why engineering is the
way it is. You will:

- Discover why engineers are different than non-
  technical people

- Understand the different branches of engineering

- Find out how to become a professional engineer
  and why it's valuable

- Learn the steps in obtaining a patent

- Expose the secret: why engineers can never
  make as much as sales and businesspeople

- Identify what motivates engineers

- Observe a simple method to solve *any* technical problem

- Pick up some useful engineering acronyms to throw out at your next big meeting

- Consider (and perhaps laugh at) quotes about engineering

- And more!

DISCLAIMER:

The author, publisher, or any named entity herein shall not be liable in any event for incidental or consequential damages from use of this material. The content of this book is completely factual, except the stuff that is not. The documented information and photo sources are accurate, but there is also subjective content. This is to say that anyone seeking to find a reason to critique will probably be successful. However, this is not intended to be a scholarly, data-based book. Rather, it is intended to be informative, educating, and (hopefully) entertaining.

So sit back, put your feet on the desk, and enjoy!

THE INSIDER'S GUIDE TO ENGINEERING

CHAPTER 1

Introduction to Engineering

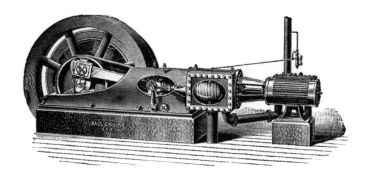

ENGINEERING, DEFINED BROADLY, IS A COMPLEX FIELD that uses math, science, and other methods to solve technical problems. In other words, engineering is the art and science by which materials and processes are made useful. The word *engineer* comes from the Latin verbs *ingenerare* (meaning to create), and *ingenium* (meaning innate quality, especially mental power or cleverness). *Engineer* dates to the 14th century when an *engine'er* was one who built or operated a siege or military engine.

Today, however, the term refers to a person who has technical training and has graduated from an engineering institute, such as a college or university.

Informally, some might describe an engineer as:

a. A person who solves a problem that was not known to exist, using methods that are not understood

b. A person who gets excited about things most others do not care about

One thing for sure: the engineer's role is big. They are required to have special understanding of technical concepts and a broad array of analytical methods, as well as in-depth knowledge of mathematics, physics, science, and technology. Supplementing this knowledge, engineers must be effective communicators in order to convey technical data to their clients, whether they be machinists, mechanics, contractors, other engineers, tradesmen, or the general public. Verbal and written communication skills are essential, since a great idea may end up going nowhere unless others can be convinced of its value. Engineers also apply established scientific principles and cutting-edge innovations to design, build, improve, and maintain complex devices, structures, systems, and processes.

The engineer may not be free to select the problem that interests them. They must solve problems as they arise, and the solution must satisfy conflicting requirements. This makes engineering a complex line of work. However, over generations, engineers have developed a simple method for solving any technical problem. Its success is guaranteed, in fact, and can easily be recalled in time of crisis. Here it is:

- *Write down the problem.*
- *Think very hard.*
- *Write down the solution.*

Works every time!

Engineers have an expansive array of problems to solve, since the field encompasses all manner of projects and demands. Engineers end up working in many different settings—including research laboratories, offices, factories, construction sites, at sea, underwater, in the air, and even in space.

You will notice this guide features many photographs of people in days gone by. As a result, almost all of the subjects are male, since female engineers did not enter the industry until the relatively recent past. This is not to suggest female engineers have no place—quite the contrary. Today, about one in five engineers is female, and even the most cursory research reveals their large impact.

The major functions of all engineering branches are as follows, in the order of most dependent on science to least:

Research:

 Using mathematical and scientific concepts, experimental techniques, and inductive reasoning, the research engineer seeks new principles and processes.

Development:

Development engineers apply the results of research to useful purposes. Creative application of new knowledge may result in a working model of some sort—such as a new electrical circuit, a chemical process, or an industrial machine.

Design:

In designing a structure or a product, the engineer selects methods, specifies materials, and determines shapes to satisfy technical requirements and to meet performance specifications. The end user is taken into consideration, as their view of the finished design is critical in the perceived value of the design.

Construction:

The construction engineer is responsible for preparing the site for work to begin (obtaining permits, construction planning, etc.), determining procedures that will safely yield the desired quality in an economical manner, directing the placement of materials, and organizing the personnel and equipment. Engineer oversight of the project can also prevent errors or accidents, if the engineer is adept and alert.

Production:

Plant layout and equipment selection are the responsibility of the production engineer, who chooses processes and provides for testing and inspection. The production engineer may observe the system in motion and advise methods to achieve higher efficiency and quality.

Operation:

The operation engineer controls machines, plants, and organizations providing power, transportation, and communication; determines procedures; and supervises personnel to obtain reliable and economic operation of complex equipment.

Management:

Project management engineers analyze customer requirements, recommend solutions to satisfy needs, and resolve related problems. They weave people, money, time, and materials into the fabric of success by achieving the necessary goals.

The fundamental concepts of the above functions are generally taught in engineering school. Students learn the basic tools to venture further into the engineering realm post-graduation.

Of all the majors available, engineering is frequently considered one of the most difficult. Why? In a nutshell, there is a lot to learn. Engineering is a very broad industry and can only be simplified so much.

Teaching a student the basics of such a wide array of disciplines demands an intense workload for a relatively short period of time. Ask someone who has been through an engineering program what it was like, and chances are that they will be more than happy to share the woes of enduring higher education.

It is all important, though. Each of the engineering functions lend themselves to a vast range of projects the engineer will face during their career.

Non-engineers often ask what engineering school is like and what skills are needed to get an engineering degree.

The foundational skill to obtain an engineering degree is aptitude in mathematics. Without this one does not get past freshman year. This is the case even though most practicing engineers do not use sophisticated mathematics directly in their work on a frequent basis. Software and calculators make complex math simpler (and much faster) for engineers today, but they still need to understand the fundamental concepts to interpret the results and apply a solution. Dependence on advanced mathematics is a dominant reason engineering is regarded as a difficult major.

Along the same lines, an interested student may wonder what the differences are in various engineering degrees: associate's, bachelor's, master's, and PhD.

The progression of degrees corresponds to increasing knowledge of the science of engineering and increasing ability to accurately predict and optimize performance prior to actual fabrication and testing of a device. An actual engineer's work is associated with the actual creation of effective devices, while the typical engineering professor is not an engineer but rather an instructor of engineering science. More succinctly, engineering professors are experts in the science of engineering.

CHAPTER 2

Disciplines
of Engineering

USC ENGINEERING PROFESSOR AND STUDENTS IN DOWNTOWN LOS ANGELES, 1912

THERE ARE MANY DISCIPLINES WITHIN THE BROADER engineering field. Each is unique in its scope of work, yet each shares the same requirement to innovate applications of natural phenomena for the use and convenience of people. Each discipline involves people, money, materials, machines, and energy, in some form or another.

The following is a list of the common engineering branches. Many interdisciplinary fields not listed here comprise of more than one of the following fields.

Acoustical Engineering

 The analysis and control of sound and vibrations. Includes noise control, room acoustics, and sound enhancement.

Aeronautical Engineering

 Design and analysis of aircraft that operate within the earth's atmosphere; heavily reliant on aerodynamics and lightweight structural efficiency.

Aerospace Engineering

 Design and analysis of satellites and spacecraft that operate beyond Earth's atmosphere.

Agricultural Engineering

 Farm machinery and agricultural structure design, natural resources, power systems, and bioenergy.

Architecture Engineering

 The art and science of engineering and construction with respect to buildings and structures.

Automotive Engineering

 The design, manufacturing, and production of vehicles. A diverse field that includes work in areas of performance, mechanical design, electrical engineering, and systems management.

Biomedical Engineering

 Applied biology, medicine, and other

life sciences for the use of advancing healthcare treatment, diagnostic devices, imaging machines, and other technology.

Chemical Engineering

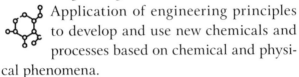 Application of engineering principles to develop and use new chemicals and processes based on chemical and physical phenomena.

Civil Engineering

One of the oldest forms of engineering; design, construction, and analysis of structures including bridges, roads, dams, buildings, etc. Includes both infrastructure development and rebuilding projects, such as after a natural disaster, for public and private sectors.

Computer Engineering

Integration of computer science and electrical engineering to develop and analyze hardware, networks, software, and computers.

Construction-Engineering Management

A specialized branch encompassing engineering and management principles to plan and administrate construction operations and projects.

Electrical Engineering

The study and application of electricity

and electronics. Involves electric and magnetic forces and their effects.

Energy Engineering

 An emerging field dealing with increasing efficiency, energy minimization, and developing renewable sources of energy.

Environmental Engineering

 The branch of engineering concerned with the environment and management of natural resources.

Forest Engineering

 The application of engineering principles to manage forest lands.

Geotechnical Engineering

 The branch of civil engineering concerned with the behavior of earth materials.

Industrial Engineering

 The design and study of logistics and industrial resources. Frequently stems from manufacturing industries, and seeks to develop efficient methods, procedures, and processes to provide continual improvement.

Manufacturing Engineering

 The design, study, and development of machines, tools, and manufacturing processes. Incorporates processes that convert materials and energy into finished products.

Marine Engineering

 The design and building of boats, ships, oil rigs, and other marine vessels.

Materials Engineering

 The development, process, and testing of materials and material properties.

Mechanical Engineering

 The application of physical principles and materials science to analyze, design, and test mechanical systems. Primary focuses are mechanics, thermodynamics, materials, and automation.

Metallurgical Engineering

 The study and development of metals and their properties.

Mining Engineering

 Application of science and technology to extract minerals from the earth.

Nuclear Engineering

 The practical application of nuclear processes—especially those used to produce and harness nuclear energy and radiation.

Petroleum Engineering

 Application of scientific principles to

detect, drill, and extract crude oil and natural gas.

Structural Engineering

 The design and analysis of load-bearing structures and supports—often a sub-discipline of civil engineering.

MECHANICAL DRAWING CLASS AT TUSKEGEE INSTITUTE, C.1902

CHAPTER 3

Good
Engineering Is...

DEFINING THE PROBLEM AND CONTINUOUSLY DRIVING TOWARD A SOLUTION

Spending the time and effort to determine the root cause of a problem usually pays dividends in finding a solution. Not knowing exactly what the problem is before attempting to solve it wastes time and effort, producing misguided research and design. Once the brass tacks are figuratively exposed, a design plan can be executed with efficiency, and one can be assured that the solution is the best one. Of course, you learn things along the way that may change the course of the project, but that's design!

ASKING YOURSELF: WOULD I PAY THIS MUCH?

A non-managing engineer may not be privy to the cost of his or her services because they do not see the invoices that their company generates. And, quite possibly, he or she does not spend much time caring about it since they are paid to be a technical person, not a businessperson. However, being informed about the costs will make the well-heeled engineer a better designer and consultant, since he or she will relate to the paying client in a more personal way. It is one of the curious surprises in life—how different things look when you are on the other side of the dollar. It is no surprise that engineering is expensive but sometimes you only get one shot at a project and it must work from the start. Enter ... the engineer.

Consider another facet: many people are not willing to pay for a perfectly optimized product. If it works, that is good enough. This attitude can be a proverbial burr under the engineer's saddle, especially when they know that a little more investment will generate a long-term solution that is far superior. However, the client is funding the project, and respect must be given to their opinion even if they disagree.

EFFECTIVELY COMMUNICATING THE SOLUTION

Engineers are at the helm of the design process and are therefore at the forefront of communicating the best solution to the receiving party. This could be the boss, another engineering team, a contractor, a non-technical client from the general public, or even a politician. If the engineer cannot get the point across by whatever means, the idea may go nowhere. Effective communication is paramount and may be one of the most challenging aspects of the engineer's responsibility.

The following image illustrates a classic example of engineering communication; engineers demonstrating how the Firth of Forth bridge works. A living model easily conveys the complexities of the structure to the non-engineering public.

Fig. 51. Living model illustrating principle of the Forth Bridge.

WORKING SUCCESSFULLY WITH OTHER PROJECT TEAMS

While some engineers would prefer to work alone, it is paramount that modern engineers obtain skills that include collaboration and teamwork with other organizations. Engineering problems today are generally too complex to be solved by a single person. The cooperation of many specialists is required. These specialists could be other engineering teams within the same company, a third-party design firm, public officials, vendors, or members of the general public.

The back-and-forth of the design process takes effort and people skills—the latter is easy to overlook for engineers. The technical work comes naturally for many engineers, but group effort, association, and partnered alliances are frequently acquired traits. The following diagram illustrates the results of group dominance, in this case the design of an airplane:

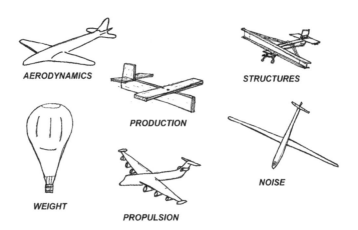

AERODYNAMICS

STRUCTURES

PRODUCTION

WEIGHT

PROPULSION

NOISE

MAKE IT IDIOT-PROOF

There is a saying about writing that is easily adapted to engineering design: *Don't engineer to be understood; engineer so that you can't be misunderstood.* It is the typical situation that shows up too frequently—the engineer can't believe the contractor doesn't know what he knows. For example, an engineer spends weeks or months analyzing a design and has the entire thing in his head. Then a contractor gets the plans and must digest the whole thing and convey it to his crew. In the process, some critical details are misinterpreted, and sure enough, it gets messed up. The engineer is frustrated that the contractor bungled such an obvious item, while the contractor is fuming at the engineer for a lousy design. You may have experienced this situation as one party or the other. One of the frequent mistakes made when designing something to be foolproof is *grossly underestimating the genius of complete fools.* There may not be a single solution to avoid costly mistakes, but the following principle at least addresses the common scenario:

ENGINEERING LAW FOR THE
CONSERVATION OF IBUPROFEN:

When an engineer intends to clearly communicate technical data, simplify the design where possible, and explain confusing details, $\lim_{n \to \infty} (x + 1)^n$ headaches can be avoided.

19

MAKING OTHER DESIGNERS FEEL LIKE FOOLS BECAUSE THEY DIDN'T THINK OF IT

"Og discovered fire, and Thorak invented the wheel. There's nothing left for us."

In engineering, there are few things more discouraging than realizing someone else came up with a better idea than yours. You don't want to feel that way. To compensate, you must come up with the better idea yourself. This can be overwhelming and difficult, to say the least, and a blank slate can be extraordinarily intimidating. However, it's a free market and the early bird gets the second mouse's cheese. Or something like that.

MAKING THE SOLUTION LOOK SIMPLE, CLEAN, AND COMPLETE

It is said that elegant simplicity is deceptively difficult to achieve. The experienced engineer knows without a doubt this is true. It takes an incredible amount of skill, knowledge, and insight to design a simple, clean solution. Beginning with this end in mind helps solve a problem efficiently.

BALANCING TIME IN THE CAD WORLD WITH TIME IN THE REAL WORLD

Contrary to an engineer's dreams, computer-aided design (CAD) is not real life. It is possible to design things in CAD that can't be built in the real world. It is also possible to design a product that *can* be built, but no one is willing to take on the job. So it is usually a good idea for the engineer to see their product from another point of view. Perhaps that means spending time operating, maintaining, and fixing the actual product. Or it could mean talking to the machinist or manufacturer about what methods work best for them before setting out to draw up plans and develop a design. Testing a prototype usually generates a wealth of insight as to what works and what does not. The concept of design for manufacturing (DFM) allows potential problems to be fixed in the design phase, and to create products that are easy to manufacture. Encouraging this kind of thinking during the CAD phase solves problems that would be too expensive to resolve later in the process.

CHALLENGING ASSUMPTIONS

Perhaps it is human nature; perhaps it is the power of habit. Or perhaps it is something else. Whatever the cause, engineers often get stuck in the same patterns—whether they be mental ruts, physical habits, or the perpetual methods of design. This is not all bad—if something clearly works, it is probably worth repeating. However, when it comes to design, innovation can be inhibited when common assumptions are not challenged. The real thought-provoking questions are obvious once they are presented, but puzzlingly difficult to recognize in the first place.

Henry Ford, in his 1926 book *Today and Tomorrow*, discusses the value of starting new operations with a person who has no previous knowledge of the subject and therefore is not on familiar terms with the problems at hand. In this way, the common assumptions of what *cannot be done* simply do not exist. The person can then ask questions—How could this not be true? How could this fail? Is this assumption actually valid?—that unearth aspects that would be otherwise missed, overlooked, or written off. Using lateral thinking, i.e., the mental process of generating ideas and solving problems by looking at a situation from a unique or creative perspective, to back out and ask the obvious questions may reveal the best solution there is.

HENRY FORD WITH MODEL T, BUFFALO, NY 1921

MAKING THE END USER HAPPY

An engineer should make the end user happy for several reasons:

- The end user may be the source of the project funding. It is probably a good idea to keep this person (or these people) happy.

- Businesspeople know that a good product sells itself. If the end user is happy with a product and the engineer's design is good, then the product sells, the business is successful, and profits increase. Pretty simple, right?

- The end user will gain respect for the engineer if the engineer spends time asking how a certain thing will be used, or what problems the user has when doing a task that relates to the product. The fact is the end user may know more about the details of a problem than the engineer does.

- By spending time with the end user, the engineer may break down barriers and create an open atmosphere for future ideas. Even if the engineer 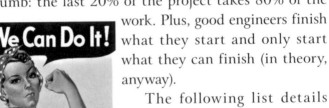 does not use the end user's ideas, just listening to their thoughts may provide context, stimulate new ideas, or make the end user feel like they are being heard.

FINISHING

An astonishingly simple goal, it is frequently one of the most difficult to complete. Finishing is the culmination of the entire engineering procedure: research, planning, design, prototyping, testing, manufacturing, and production. Consequently, it may be considered the most important step in the process. Remember, an unfinished machine *never* works. The 80/20 rule is a good rule of thumb: the last 20% of the project takes 80% of the work. Plus, good engineers finish what they start and only start what they can finish (in theory, anyway).

We Can Do It!

The following list details some qualities of both good and bad engineers. These categories are subjective, to a degree (i.e., a certain project or situation may demand a quality that is normally considered "bad"). But the idea here is to distinguish the differences, and with a bit of luck, inspire the reader to migrate toward the more desirable side.

GOOD ENGINEERS	BAD ENGINEERS
Take initiative	Wait for instructions
Continuously improve	Are complacent
Are creative	Are chaotic
Ask questions	Are afraid of looking dumb
Are disciplined	Are unpredictable
Answer questions	Get offended
Put the team first	Are in it for themselves
Learn from failure	Deny that failure occurred
Build relationships	Build a wall (reinforced masonry)
Build what the customer needs	Build what the customer asks for
Get to know their clients	Disdain their clients
Recognize other good engineers	Want all the recognition
Are constantly learning	Solve new problems with old methods
Teach other engineers	Are black holes
Own the problem and solution	Point out problems
Are leaders	Think they are leaders
Make other engineers better	Keep success to themselves
Make things happen	Wonder what happened
Systematize	Disorganize
Make logical assumptions	Guess
Adjust quickly and are flexible	Refuse to adapt
Prioritize	Compromise
Are consistent	Are erratic

CHAPTER 4

Engineering Licensure

ACCORDING TO THE NATIONAL COUNCIL OF EXAMINERS FOR ENGINEERING & SURVEYING (NCEES), professional engineering licensure protects the public by enforcing standards that restrict practice to qualified individuals who have met specific qualifications in education, work experience, and exams. Engineering licensure indicates dedication, leadership, and advanced management skills. In the United States, each state regulates licensure for the engineering and surveying professions. Candidates interested in pursuing licensure are encouraged to check the requirements in the state or territory where they plan to practice, as the requirements vary.

The Fundamentals of Engineering (FE) exam, also

known as the Engineer in Training (EIT) exam, is generally the first step in the process to becoming a licensed professional engineer (PE). It is designed for recent graduates and students who are close to finishing an undergraduate engineering degree from an ABET-accredited program.* Some states permit students to take the exam prior to their final year in school. The FE exam covers topics ranging from general engineering subjects to math and science fundamentals. Once a person passes the FE exam, they become an Engineering Intern (EI).

The Principles and Practice of Engineering (PE) exam tests for a minimum level of competency in an engineering discipline. It is designed for engineers who have gained a minimum of four years' post-college work experience in their chosen engineering discipline.

Exams are discipline-specific, meaning the examinee selects the exam that most suits their experience and education.

Upon successfully passing the FE exam, the PE exam, and completing four years of approved engineering work, the applicant will receive a professional engineering license from their state board. Some states allow EITs to take the PE exam before their experience is complete.

Completing the licensure requirements permits the licensed engineer to obtain their own stamp, and

---

* ABET is a global organization that focuses on approving and authorizing educational and technical programs within colleges and universities.

subsequently seal plans, designs, documents, or other technical material of which he or she has responsible charge and oversight. Stamping a drawing or document is an official act that ties the engineer of record to the design. Both ink stamps (a.k.a. "wet signatures") and digital stamps are commonly utilized. Doing engineering work in another state requires licensure in that state. Once you have a license in one state, you can apply by *comity* or *reciprocity* to get as many additional licenses as you need. Certain states require extra training to become licensed in their state.

Some engineers become structural engineers (SE) in addition to being a PE. Further experience and examination are required to get a structural engineering designation.

The following figures show examples of professional engineering stamps from various US states and territories.

People who want to start their own business, or inquire about professional engineering and its services, often ask the following questions:

WHY SHOULD I HIRE A PE?

Professional engineers have specific expertise in certain subjects. Each PE has taken the exam pertaining to his or her field of work, and they should be proficient within that sector. If your business plan includes a focus on, say, the design of composite panels for portable machines, a mechanical engineer with experience in mechanical systems and materials would be a great asset. If you plan to develop a new method to identify heavy metals in fruit juice, a chemical engineer is the person you need. Advanced development for high-tech industries rely heavily on electrical and software engineers, among others. There are many examples of situations where a specialized engineer is the perfect solution.

WILL HIRING OR CONTRACTING WITH A PE HELP ME START MY OWN BUSINESS?

Probably. The value of specific expertise has already been stated. Along with that, a PE carries the liability for their design, meaning that your business responsibility is less since the engineer owns the risk and personal liability of their work. Plus, most if not all states require a person to be licensed to be a consulting engineer, or practice engineering services for the general public. Exemptions to this are generally engineers employed by manufacturing companies—whose engineering services are only provided within the company, and not to the general public.

WHY MUST I HIRE A PE?

If a design for a certain project will be submitted to local authorities, such as state or county officials, it will be subject to the current structural or building codes within the particular jurisdiction. Permits granted by government authorities usually require calculations and design drawings that are stamped by a licensed professional. However, there are also allowances for designs that meet code criteria and do not need to be stamped. For example, a plans examiner may allow you to design your own residential home if it meets the minimums specified in the local code, and a PE stamp is not required. Projects that are not within the "standard" realm of the applicable code will likely necessitate the involvement of a PE.

Another benefit of having a PE on the payroll is to help obtain funding for a certain project. In fact, it may be required. Recall the Golden Rule of Business: the person with the gold makes the rules.

WHAT SHOULD I LOOK FOR WHEN HIRING A PE?

- Make sure the PE is in good standing with your state board of engineering

- Confirm the PE holds a license in your state

- Check the PE's record with the state board; have they been fined for unethical or criminal behavior?

- Verify the validity of their firm and business history

- Ask around—talk to contractors and other engineers

  - Is this PE easy to work with?

  - Are they flexible?

  - Do they listen to the customer's needs?

  - Do they work efficiently?

  - How much do they charge? (Although you should not hire an engineer solely on their fee)

  - Do they conduct themselves in a professional manner?

  - Do they possess a good character?

- See if they carry something to write with and something to write on. This is one subtle indicator of a good engineer.

- Notice if the engineer draws you a picture while

explaining a design. It is a big bonus if you can tell what they are drawing.*

HOW MUCH DOES A PE CHARGE?

It depends. Factors that influence an engineer's fee include location (which state or region the work is being performed in), the PE's experience level, availability, type of industry, etc. Fees can vary greatly—and it depends on how much they want to do your project and what their incentive is. A PE may not be licensed in your state, and if they have to apply for and obtain an additional license, this adds costs and time.

HOW DOES A PE CHARGE FOR THEIR WORK?

Engineers have several options on how to charge for their services. The following list presents the standard methods:

> *Lump Sum Fee*: This is a predetermined fee agreed upon by the client and engineer. Usually better for small projects where the scope of work is easily defined.

> *Per Diem Fee:* The engineer is paid a certain sum

---

* For unknown reasons, it is virtually impossible for an engineer to describe structures or mechanical principles without using their hands or drawing a picture.

for each working day. Some expenses may be billed in addition to the per diem amount.

Salary Plus Fee: The engineer is paid as salaried plus a percentage for overhead and profit, and other expenses.

Cost Plus Fixed Fee: All costs incurred by the engineer are paid by the client in addition to a predetermined fee as profit. Best used when the scope is difficult to determine in advance. Good records must be kept with this method.

Retainer: A non-refundable minimum amount paid by the client to the engineer in advance. Think of it as a down payment for an engineer to do your job.

Percentage of Construction Cost: The engineer is paid a certain percentage of the total project construction cost. Some fees, such as legal fees, project management labor, plan revisions, etc. may not be included in the total project cost.

HOW CAN HIRING A PE PROTECT MY BUSINESS?

Hiring a PE shifts the liability to the engineer, rather than your business. PE licenses are granted to individuals, not businesses, so even though a company has responsibility of the entire project, the PE holds the responsibility of the design. Also, by hiring a PE, you are hiring a licensed, qualified expert and hopefully "dumb" mistakes are avoided in the first place.

WHY IS A PE DIFFERENT THAN OTHER PROFESSIONALS?

Primarily, a professional engineer has advanced technical knowledge of their sector of engineering and has passed exams that demonstrate minimum competence in a certain discipline.

Second, the PE has a stamp, which many other professionals do not employ. Exceptions to this are architects, professional land surveyors, geologists, and, curiously, interior designers—among others. The stamp links the PE to the design as they become the engineer on record, making it difficult to separate themselves from an approved design or document. Contrast this with professions that do not use a stamp—a salesman or a clerk, for instance—where it might be easy to slip out of a bad deal if your work lacks a stamp, license number, or signature.

PEs are also held to a high standard of ethics and professional expectations. The fundamental rules of engineering are:

- Hold paramount the safety, health, and welfare of the public.

- Perform services only in areas of competence.

- Issue public statements only in an objective and truthful manner.

- Act for each employer or client as faithful agents or trustees.

- Avoid deceptive acts.

- Be honorable, responsible, ethical, and lawful, so as to enhance the honor, reputation, and usefulness of the profession.

Frequently, an engineer's work results in a life-protecting or safety-critical design: for example, a bridge, vehicle brakes, elevators, aircraft, electrical hardware, water treatment systems—the list goes on. Each of these could be potentially dangerous or fatal for the general public if subject to catastrophic failure. This makes the engineer's job and responsibility very significant.

Herbert Hoover, the thirty-first US president and a former engineer, wrote that:

[Engineering] is a great profession. There is the fascination of watching a figment of the imagination emerge through the aid of science to a plan on paper. Then it moves to realization in stone or metal or energy. Then it brings jobs and homes to men. Then it elevates the standards of living and adds to the comforts of life. That is the engineer's *high privilege*.

The great liability of the engineer compared

to men of other professions is that his works are out in the open, where all can see them. His acts, step by step, are in hard substance. He cannot bury his mistakes in the grave, like the doctors. He cannot argue them into thin air, or blame the judge, like the lawyers. He cannot, like the architects, cover his failures with trees and vines. He cannot, like the politicians, screen his shortcomings by blaming his opponents and hope the people will forget. *The engineer simply cannot deny he did it ...*

... On the other hand, unlike the doctor, his is not a life among the weak. Unlike the soldier, destruction is not his purpose. Unlike the lawyer, quarrels are not his daily bread. To the engineer falls the job of clothing the bare bones of science with life, comfort, and hope. No doubt as years go by the people forget which engineer did it, even if they ever knew. Or some politician puts his name on it. Or they credit it to some promoter who used other people's money ... But the engineer himself looks back at the unending stream of goodness which flows from his successes with satisfactions that few professions may know. And the verdict of his fellow professionals is all the accolade he wants.

SHOULD I BECOME A PE?

It depends. If you have an interest in becoming a consulting engineer, working in government engineering positions, or establishing a private engineering practice, then yes, you should become a PE. In many of those cases, it is not only a desirable status but a legal requirement. A career in engineering education may find a PE license valuable. If you have no interest in developing your engineering career, have no current need to become a PE, and are willing to pass up future engineering job opportunities, you should probably focus your time and effort in another endeavor.

WHY AREN'T ALL ENGINEERS LICENSED?

There are several reasons. For one thing, it takes work to get a PE license, as mentioned before. And it takes time and money to take the exam, pay the recurring license fees, and maintain continuing education requirements. Many engineers work in an industry where a PE isn't required or even useful. Consequently, the investment isn't worth it to them.

CAN I BE AN ENGINEER WITHOUT A HAVING PE LICENSE?

It depends. Some argue that this is a moot point—if you graduated from an engineering college with an engineering degree, if your business card says "engineer," if you work in the engineering department, and if you do engineering work, then of course you are an engineer! But not so, say some state boards. Technically, many states and local governments do not consider an individual to be an engineer unless they are a PE. Often, the divide in this controversy is the difference between work in the public sector and the manufacturing industry. In manufacturing, generally speaking, you can be an engineer without holding a PE license. In public sector work (projects submitted or processed by a government entity), you have to be a PE to be considered an engineer. As you might imagine, this can be a touchy subject.

Another variation on this issue is a tradesperson who is good at building things that work. Can they be called an engineer? The difference here is the ability an engineer has to accurately predict and optimize performance *prior* to fabrication and testing of a device.

WHAT ARE SOME OF THE PITFALLS OF THE INTERACTION BETWEEN ENGINEERS AND THE GENERAL PUBLIC?

Engineers frequently work on things that are important but not obvious, like foundations and utilities. Or they work to mitigate problems that may never happen—such as fire, severe weather, and earthquakes. Although vital to public safety, these types of projects may be perceived by the public as necessary but not interesting. Many people would rather pay for fancy drapes or a piece of fine art than a properly designed footing.

Also, engineers are typically invisible to the general public. They do not normally hold celebrity status. Engineers do not usually run for high profile elected positions (there are always exceptions, of course). This is not to say engineers are hermits; quite the contrary.

But the public easily overlooks engineers and the critical value they add to the efficiency, practicality, and safety of everyday life. There are so many larger-than-life figures who make news that is more interesting to the common man. Other professionals—doctors, lawyers, plumbers, and so on—solve problems once they become apparent, and they are recognized for it. Engineers solve problems *before they become problems*, and their work is frequently not recognized or even known.

WILLIAM H. BOWLUS (FACTORY MANAGER), B. FRANKLIN MAHONEY (PRESIDENT), CHARLES LINDBERGH, DONALD A. HALL (CHIEF ENGINEER AND DESIGNER), AND A.J. EDWARDS SALTES (MANAGER) IN FRONT OF THE *SPIRIT OF ST. LOUIS* AT RYAN AIRCRAFT CO. IN SAN DIEGO, MAY 1927.

CHAPTER 5

Patents

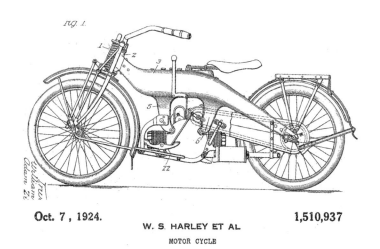

rug. 1.

Oct. 7 , 1924.   1,510,937

W. S. HARLEY ET AL

MOTOR CYCLE

W HAT IS A PATENT? IN A NUTSHELL, IT IS A legal right and a government-granted monopoly to be the only one who can make, use, or sell an invention for a certain number of years. Protecting an invention may also require a trademark, copyright, marketing plan, or trade secret. To obtain a patent, an invention must be novel, non-obvious, adequately described, and claimed by the inventor in clear and definite terms. This short summary addresses some of the questions that engineers frequently ask when dealing with intellectual property.

WHAT ARE THE DIFFERENT KINDS OF PATENTS?

Utility patents are granted to applicants who invent or discover a new and useful process, machine, article of manufacture, or composition of matter—or any new useful improvement thereof. Most US Patent and Trademark Office (USPTO) patents are utility applications.

Design patents are granted to applicants who invent a new, original, and ornamental design for an article of manufacture.

Plant patents are granted to anyone who invents or discovers and reproduces any distinct and new variety of plant.

WHY ARE PATENTS NECESSARY?

Patents provide an incentive for companies to innovate and share their ideas with the public. In exchange for disclosing an invention, the government will grant a limited monopoly to the patent owner.

WHAT CAN'T BE PATENTED?

- laws of nature

- physical phenomena

- abstract ideas

- literary, dramatic, musical, and artistic works (although these can be copyright-protected)

- inventions which are:

  - not useful (such as perpetual motion machines)

  - offensive to public morality

WHAT ARE THE STEPS TO OBTAINING A PATENT?

To obtain a patent, you must file a patent application with the USPTO. Foreign countries have similar counterparts. Once the application is filed, it is examined by the USPTO and there are generally a number of iterations before the patent applicant receives a Notice of Allowance. Once the application is allowed, the applicant pays an issue fee and the patent will be granted soon thereafter.

The USPTO promotes the use of an eight-step patent process:

1. Determine the type of intellectual property protection that you need

2. Determine if your invention is patentable

3. Determine the kind of patent you need

4. Get ready to apply

5. Prepare and submit your initial application

6. Work with your examiner

7. Receive your approval

8. Maintain your patent

WHAT HAPPENS AFTER I RECEIVE A PATENT?

When you receive a patent, you are granted a patent number by the USPTO, along with a copy of the issued patent. Periodic maintenance fees are required to maintain the patent in force. Ultimately, the patent expires—generally twenty years after the earliest filing date of a corresponding patent application.

CAN OR SHOULD I SELL MY PATENT?

Yes, you can sell your patent. Whether you should sell it, however, is quite another matter. Selling your patent is a business decision that is influenced by several different factors. There are three general ways to obtain tangible financial benefit for a patent:

- *Sale*
- *Licensing*
- *Exercising market power*

If a company is no longer using patented technology, often the company will choose to sell the patent.

BEFORE YOU APPLY ...

Conduct a search to identify any prior filings that describe features related to your invention. After the search, you are better prepared to draft an application that will provide proper scope for your invention. The application scope should not be too narrow or too broad.

ARE THERE VALID INFORMAL WAYS OF GETTING A PATENT?

No. Patents must be obtained from the USPTO through the application process.

RESOURCES FOR PATENT APPLICANTS:

1. General Information: www.uspto.gov/patent

2. www.google.com/patent

3. http://patft.uspto.gov/

4. International Information: https://www.wipo.int/portal/en/

5. https://patentlyo.com/

6. https://www.ipwatchdog.com/

7. *Patent it Yourself* (D. Pressman, 2002)

8. *Patents Demystified* (D. Adams, 2015)

9. *Fun with Patents* (K. Luzzatto, 2016)

10. *Navigating the Patent System* (J. Yang, 2017)

CHARLES W. OLDREIVE'S "NEW IRON HORSE", C. 1882

CHAPTER 6

Managers, Salary, & Non-Technical People

In a 1912 issue of *Flight* magazine, British inventor W.T. Warren's invention, a protective flight helmet, is demonstrated. The image is often erroneously reported to be a football helmet. The man in the photo does not play football. The wall, against which the helmet carrier ran, belongs to a hangar of the flying school of William Hugh Ewen. The owner himself (middle back) and his chief pilot (back left) are in the photo. The fellow in the foreground is his student W.T. Warren. And, no, he is not reacting in anger because he failed his flight test. The leather cap presented by Warren was padded with horsehair; a system of steel springs should intercept any impact, thus reducing the risk of injury. Head injuries were the leading cause of death in flight accidents.

AS AN ENGINEER, ONE MAY FIND IT CHALlenging to follow the lead of a manager—especially when the manager is not an engineer. Circumstances involving a manager may drive you to literally bang your head against the wall—which, in fact, is bad for the wall. The following story illustrates this sentiment:

> A man is flying in a hot air balloon and he realizes that he is lost. He reduces altitude and spots a man down below. He lowers the balloon further and shouts, "Excuse me. Can you help me? I promised my friend I would meet him a half hour ago, but I don't know where I am."
>
> The man below says, "Yes. You are in a hot air balloon, approximately 30 feet above this field. You are between 45° and 46° N latitude, and between 120° and 122° W longitude."
>
> "You must be an engineer," says the balloonist.
>
> "I am," replies the man. "How did you know?"
>
> "Well," says the balloonist, "everything you have told me is technically correct, but I have no idea what to make of your information, and the fact is I am still lost."
>
> The man below says, "You must be a manager."
>
> "I am," replies the balloonist. "How did you know?"

"Well," says the man below, "you don't know where you are, or where you are going. You have made a promise that you have no idea how to keep, and you expect me to solve your problem. The fact is you are in exactly the same position you were in before we met, *but now it is somehow my fault.*"

However comical this example may be, it is imperative that modern engineers develop verbal and written communication and other "soft" skills that enable them to work across department lines. For example, a consulting engineer may frequently work directly with the general public or non-technical people. Often, these situations call for the engineer to exhibit extra sensitivity and forethought, as non-engineers may not be interested or may not understand the level of detail an engineer is attempting to convey.

Another example: James Thinker, PE, and his engineering firm Thinker & Associates, Inc. spend significant time and effort developing a robust design for a new public structure. Then, for a reason not associated with the design, a visible failure occurs and a stream of public critiques flow to Thinker and his engineers, who must wade through the fray and respond in a professional manner. This type of public scrutiny on a frequent basis may well be the undoing of a normally tough-minded engineer. An adapted version of Newton's Third Law of Motion applies to these types of circumstances: *For every action, there is an equal and opposite criticism.*

Along the line of engineers being managed, there is a relationship that is worthy of special attention. That is the interaction between architects and engineers. Typically, architects are concerned with many things, including space distribution, construction, financing, human behavior and tendencies, aesthetics, real estate, and so on. Engineers, however, are much more focused, specializing in certain areas of expertise that pertain to a structure and the forces upon it. No doubt the differences in each are bound to conflict. As Dr. Mario Salvadori says in his book *Why Buildings Stand Up*, "Lucky is the client whose architect understands structure and whose structural engineer appreciates the aesthetics of architecture." Although drafting is said to be the chief form of engineering communication, the dialogue between architect and engineer usually goes far beyond graphical exchange.

In the engineering business (and other businesses for that matter), there are three steps to success. One is finding the work. Two is doing the work. Three is getting paid for the work. As a young engineer, or student, you are only aware of a fraction of step two. However, all three steps are of vital importance to stay in business. Experience and advancement open an engineer's eyes to business features that were previously unknown or disregarded.

Regarding salary, engineers receive relatively good compensation. Clients usually recognize the value of having a project engineered to reduce risk and develop detailed plans, and the consulting engineer can capture this benefit albeit in the duty to uphold public safety in an ethical way. Of course, engineering is frequently expensive and can make up a substantial part of a project's budget. A good engineer keeps the client's costs in mind and their work should add great value to the project.

However, there is a notion that engineers can never earn as much as business executives and salespeople. *Dilbert's Salary Theorem* categorically proves this. Mathematical equations based on the following two postulates support this theorem:

1. Knowledge is power

2. Time is money

As every engineer should know:

$$\text{power} = \frac{\text{work}}{\text{time}}$$

Substituting in the two postulates:

$$\text{knowledge} = \frac{\text{work}}{\text{money}}$$

Solving for money, we get:

$$\text{money} = \frac{\text{work}}{\text{knowledge}}$$

Notice that as knowledge approaches zero, money reaches infinity regardless of the amount of work done.

Conclusion: The less you know, the more you make.

As true as this may be, the author does not intend to discourage anyone from the field of engineering by insinuating educated engineers have a financial disad-vantage. Rather, consider this an explanation of why the world is the way it is. Engineers should recognize the work they do is not strictly for compensation but for the benefit of their company, the good of the general public, and the satisfaction of a job well done (which may be payment enough).

Sometimes non-engineers have trouble understanding what engineers are saying. And sometimes this is intentional on the engineer's part. It may be beneficial to not inform everyone about everything. Here are a few semi-valuable phrases and their hidden meanings engineers may employ:

"A number of different approaches are being tested and the best one will be selected once the results are conclusive."

We have no idea what's going on and we are trying things at random to see what happens.

"We'll generate successful progress reports as things develop."

Hopefully, something works soon.

"The test results were satisfactory."

It didn't blow up and, boy, were we surprised!

"Teamwork is essential on this project; we need everyone's input."

If this project goes bad, we want all of you to bear the blame.

"Years of intensive research have revealed the perfect solution."
    We tried everything we could think of and just happened to recently stumble onto the answer.

"The final design is pending because its development is ongoing."
    We haven't started yet.

"Our conclusion is that the problem is unsolvable with the resources we currently have available."
    This is a lot tougher problem that we thought.

"This project will be postponed until further resources are available."
    The only guy who knows about this just quit and moved to another country.

CHAPTER 7

The
Engineer
Personality

WHILE SOME PERSONALITIES THRIVE ON SOCIAL interfacing, frequent meetings*, and group activities, engineers usually do not. They see things black and white with few other options (of course, this is painting with a broad brush—needless to say, all engineers are not the same). However, a stereotype seems to shadow the engineer, and it may in fact be accurate.

In 1954, Dr. Charles E. Goshen, a diplomate of the American Boards of Psychiatry and Neurology, authored an essay titled "The Engineer Personality," which makes a strong case that certain personality characteristics are

---

*   See Appendix C

attached to various professions, specifically within engineering. The article points out that:

> The engineering profession is one area of human endeavor wherein there is a very high consistency insofar as the character traits in which its members have in common are concerned. In other words, *there exists an "engineer personality"*. As with all other typical personalities, in various occupations, it must be kept in mind that it is not the occupation itself which determines the personality, but rather it is the type of personality which chooses the occupation in question. In other words, *the engineer already had his particular set of personality traits before he became an engineer.*

For some engineers, this may be a huge relief, since it explains a lot if you ever wonder, "How did I get this way?" Now you know the truth, and a deep sigh is quite in order.

The same article points out that the engineer's most obvious characteristic is his "precision, his meticulousness, his attention to detail and accuracy or his perfectionism." And "once we get to know an engineer better, we appreciate that his intelligence tends to be used in a very specialized way ... His success in mastering mechanics tends to lead him farther away from achieving competence in dealing with people."

Dr. Goshen concludes that because of this confidence with mechanics, engineers tend to apply the same principles to people, with the inevitable failure in achieving a successful relationship. Also, he clarifies that engineers exhibit an enormous need to be right. Looking into this more closely:

> We find he is primarily interested in trying to avoid being criticized for being wrong. As a result, he demonstrates an outstanding sensitivity to criticism.

"I Married an Engineer," an article published in 1973, states that:

> One wife believes there are two kinds of engineers: those who "could almost have chosen any other means of livelihood" and "the true engineer." The true engineer, she says, is "a logical, rational being, who can usually evaluate a problem, situation, or person precisely. He deals in truths, in facts, and sets high standards of performance for himself and everyone or everything connected with him. He has great integrity in his personal relationships, thinks for himself in politics, current affairs. He would appear stiff and old-fashioned on initial meeting, because he usually does not make small talk. But upon closer acquaintance, he can be found to be a well-informed, interesting person."
>
> Another wife describes the engineer as a "quiet perfectionist, terribly practical and economical, very level-headed." Still another agrees, "very quiet. At times, hard to converse with. Well-read, and when he wants to talk, he can talk on almost any subject."

The wife of an aerospace engineer says the engineer "uses step-by-step logical, rather than impulsive, reasoning. Expresses himself extremely well, but with great forethought and deliberation. Conceals deep emotions."

Engineers are indeed handy fellows at home, for both improvement projects and repairs. Moreover, their analytical minds will not only find solutions for existing problems but also sort out precisely what the problem is. For instance, one wife was struggling to cut a piece of carpeting into a runner to cover the children's traffic pattern into the kitchen. The engineer in the family quickly saw that the problem was not how to cut the carpeting but how to change the traffic pattern. And he found a simple way to do it.

For non-engineers, without the aforementioned knowledge, it may be tough to understand why engineers act the way they do. This is a common perception when observing anyone with whom you have not taken time to get to know. You do not know what that person happens to be dealing with. For those in engineering, the loads can be intensely heavy. Every force applied to the individual is felt and is ideally balanced by an equal and opposite force reacting in the opposite direction. Both engineering students and career engineers experience similar conditions. Free-body diagrams demonstrate this effect:

Engineering Student

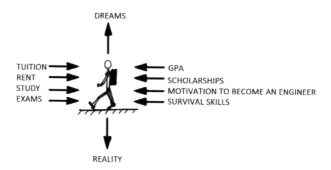

As the student transitions into a full-blooded engineer, the applied forces still exist, but shift to heavier loads:

Engineer

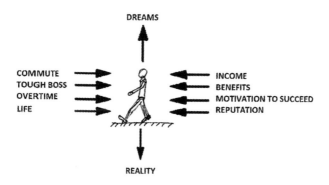

The non-engineer can benefit from this representation by realizing that the engineer experiences the same resistance as most other professionals. However, engineers carry greater responsibility than many and often have to solve a full plate of technical and advanced problems that go far beyond what the non-engineer probably realizes.

The true value of an engineer is subjective, to a degree. Anyone can (but probably will not) read a salary survey that points out their value if they are so-and-so in a certain quartile of a particular state. But what if you are a department director who just wants to get a qualitative idea of how valuable your hired engineer is? Say no more! This comprehensive list is all you need:

Engineering Hierarchy and Relative Value

Retired Engineer	♦♦♦♦♦♦♦
Principal Engineer	♦♦♦♦♦♦
Director of Engineering	♦♦♦♦♦
Chief Engineer	♦♦♦♦♦
Engineering Manager	♦♦♦♦
Senior Engineer	♦♦♦
Project Engineer	♦♦
Engineer in Training	1 ½ ♦
Engineering Intern	½ ♦
Engineering Student	-♦♦♦*

Regarding social interaction, engineers have different objectives than other people. These objectives may develop over time or be the direct result of a singularly poor social experience as a child, which consequently may contribute to the forging of a great engineer. Referring to the article by Dr. Goshen again, the engineer's success in mastering mechanical principles leads him or her to apply the same principles to people with inevitable failure in successful relationships with people. Physical

---

* Possibly -♦♦♦, depending on circumstances.

laws and their predictability lead them to expect the same kind of predictability in people.

The following text is a summary describing the engineer's perspective on social events:

Normal people expect to accomplish several unrealistic things from social interaction:

- Stimulating and thought-provoking conversation

- Important social contacts

- A feeling of connectedness with other humans

Engineers, however, have rational objectives for social interactions:

- Get it over with as soon as possible

- Avoid getting invited to something unpleasant

- Demonstrate familiarity and knowledge of various topics

On another note, engineers are frugal. This is not because of cheapness or a mean spirit. It is because every spending situation is simply a problem in optimization—that is, "How can I escape this situation while retaining the greatest amount of cash?"*

The fastest way to get an engineer to solve a problem is to declare that the problem is unsolvable. No engineer can walk away from an unsolvable problem until it is solved. No illness or distraction is enough to get the engineer off the case. These types of challenges quickly become personal—a battle between the engineer and the laws of nature.

Nothing is more threatening to the engineer than the suggestion that somebody has more technical skill. Normal people sometimes use that knowledge as a lever to extract more work from the engineer. When an engineer says that something can't be done (a code phrase that means it's not fun to do), some clever normal people have learned to glance at them with a look of compassion and pity and say something along these lines: "I'll ask Bob to figure it out. He knows how to solve difficult technical problems."

At that point it is a good idea for the normal person to not stand between the engineer and the problem. The engineer will set upon the problem like a starved dog on a pork chop.

---

* Engineers are probably as economical with their time as with their money. Characteristics like hyper-efficiency in an engineer can be misinterpreted by non-engineers as being unusual or eccentric to the point of being comical, largely at the expense of the engineer.

CHAPTER 8

Engineering Acronyms

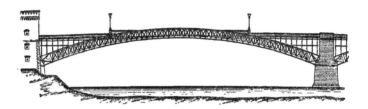

THE FOLLOWING LIST IS A SAMPLING OF ACRONYMS and abbreviations that an engineer may encounter in various projects or drawings. Some of these terms are 24K gold nuggets that will impress the boss and colleagues at your next big meeting. Few things say "This guy is amazing" like rattling off something about the company's great STEP program and how the Standard for the Exchange of Product Model Data is meeting engineering department goals. Or perhaps you can easily recall off the top of your head what UHMW stands for when a client calls to verify the materials list. Some of the other acronyms here may only find value in Christmas party small talk with other engineers. Discussions with a non-technical person (NTP)

about engineering, especially when using abbreviations, is often one-sided and interspersed with blank stares and confused looks, and it is well-advised to use acronyms with discretion in those circumstances.

Regardless of the situation, anyone seeking a greater understanding of engineering acronyms is recommended to read this list:

ABET	Accreditation Board for Engineering & Technology, Inc.
AC	Advisory Circular
	Asphaltic Concrete
	Alternating Current
AG	Above Grade
AI	Artificial Intelligence
AISI	American Iron and Steel Institute
AL	Aluminum
AMS	Aerospace Material Standards
ASD	Allowable Strength Design
ASTM	American Society for Testing and Materials
A/R	As Required
AWG	American Wire Gauge
BC	Bolt Circle
BG	Below Grade
BHN	Brinell Hardness Number
BM	Benchmark
BMP	Best Management Practice
BOM	Bill of Materials
CAD	Computer Aided Design/Drafting
CAGE	Commercial and Government Entity
CG	Center of Gravity

CIP	Cast in Place
	Cast Iron Pipe
CL	Center Line
CMP	Corrugated Metal Pipe
CNC	Computer Numerical Control
CP	Control Point
CRS	Cold Rolled Steel
CBORE	Counterbore
CSINK	Countersink
CSP	Corrugate Steel Pipe
DC	Direct Current
DFM	Design for Manufacturing
DIA	Diameter
DIP	Ductile Iron Pipe
DL	Dead Load
DOE	Design of Experiments
DWG	Drawing
DXF	Data Exchange Format
ED	Edge Distance
ECO	Engineering Change Order
ECN	Engineering Change Notice
EDM	Electro-Discharge Machining
EMF	Electromotive Force
EXIST	Existing
FAR	Failure Analysis Report
FEA	Finite Element Analysis
FL	Floor Level
	Flow Line
FN	Flag Note

FRACAS	Failure Reporting, Analysis, and Corrective Action System
FS	Far Side
	Factor of Safety
GD&T	Geometric Dimensioning & Tolerancing
GIS	Geographic Information System
HC	Horizontal Curve
HDPE	High Density Polyethylene
HOI	Height of Instrument
HR	Hardness, Rockwell (A, B, C)
HRS	Hot Rolled Steel
HT TR	Heat Treatment
H&T	Hardened and Tempered
IAW	In Accordance With
ID	Inner Diameter
IE	Invert Elevation
IM	Iron Monument
IP	Iron Pipe
JIT	Just-In-Time Manufacturing
KIP	1000 lb.
KSI	KIPs per square inch
LiDAR	Light Detection and Ranging
LL	Live Load
LMC	Least Material Condition
LRFD	Load and Resistance Factor Design
MEK	Methyl Ethyl Ketone
MF	Make From
MFG	Manufacturing
MH	Manhole
ML	Main Line

MMC	Maximum Material Condition
MP	Milepost
MRB	Material Review Board
MS	Margin of Safety
MSDS	Material Safety Data Sheets
MTBF	Mean Time between Failures
NA	Neutral Axis
	Not Applicable
NC	Numerical Control
NCEES	National Council of Examiners for Engineering and Surveying
NCM	Nonconforming Material
NCR	Nonconformance Report
NEC	Not Elsewhere Classified
	National Electric Code
NOM	Nominal
NORM	Normalized
NPT	National Pipe Taper (Thread)
NSPE	National Society of Professional Engineers
N&T	Normalized and Tempered
NTS	Not to Scale
OAL	Overall Length
OC	On Center
OD	Outer Diameter
OHL	Over High Limit
ORIG	Original
PC	Precast
	Point on Curve
PCB	Printed Circuit Board
PCC	Portland Cement Concrete

PDH	Professional Development Hour
PDM	Product Data Management
PI	Point of Intersection
PL	Parts List
	Pipeline
PLM	Product Lifecycle Management
PN	Part Number
POI	Point of Intersection
POT	Point on Tangent
PSI	Pounds per Square Inch
QMS	Quality Management System
QTY	Quantity
R	Radius
RCP	Reinforced Concrete Pipe
REF	Reference
REQ'D	Required
REV	Revision
RF	Radio Frequency
RFP	Request for Proposal
RFQ	Request for Quote
RM	Reference Monument
RP	Reference Point
RMS	Root Mean Square
RT	Room Temperature
	Rough Turned
RTP	Release to Production
R/W	Right of Way
RZ	Roughness

SD	Sight Distance
	Storm Drain
SE	Super Elevation
SG	Specific Gravity
SFACE	Spot Face
SHN	Shown
SN or S/N	Serial Number
SPP	Structural Plate Pipe
SPPA	Structural Plate Pipe Arch
SS	Stainless Steel
	Supersede
STD	Standard
STEP	Standard for the Exchange of Product Model Data
STL	Steel
TB	Title Block
TCE	Thermal Coefficient of Expansion
THD	Thread
THK	Thickness
THRU	Through
TOL	Tolerance
TYP	Typical
UAI	Use As-Is
UHF	Ultra High Frequency
UHMW	Ultra High Molecular Weight (Polyethylene)
ULL	Under Low Limit
UNC	Unified National Coarse (Thread Standard)
UNEF	Unified National Extra Fine (Thread Standard)
UNJC	Unified National J Series Course (Thread Standard)

UNJF	Unified National J Series Fine (Thread Standard)
UOS	Unless Otherwise Specified
UTS	Ultimate Tensile Strength
VC	Vertical Curve
VHF	Very High Frequency
WBS	Work Breakdown Structure
WGS	World Geodetic System
WI	Wrought Iron
	Within
WIP	Work in Progress
WW	Water Well
YS	Yield Strength
YGWYPF	You Get What You Pay For

CHAPTER 9

Conclusion

BOEING-VERTOL 107-II CERTIFICATION TEAM INCLUDING MANUFACTURING MANAGER, QUALITY CONTROL MANAGER, PROJECT ENGINEER, THE FLIGHT TEST ENGINEERING MANAGER, THREE TEST ENGINEERS, TWO BOEING PILOTS, TWO FAA PILOTS, AN FAA INSPECTOR, AND TWO FAA ENGINEERS.

ENGINEERING IS A GREAT INDUSTRY. WIDELY DIVERSE, universally applicable, and technically challenging, engineering provides interesting projects and valuable developments to the global market. Its rich history and past projects are useful references for modern engineers who create innovative solutions. And there is no end to the problems that need to be solved.

Over time, engineering has changed. The photos in this book are evidence of this, featuring an antiquated methodology and culture that is much different today. The advent of the digital age has also impacted engineering in a major way. One specific example of this is the loss of skill engineers used to have for pencil-in-hand artistry. There was a time when being able to create accurate pencil-and-paper drawings was a key skill for both engineers and mechanical designers. This aspect of the field has almost entirely disappeared. Software and computers add great efficiency and ability to the modern engineer, but at the cost of other things going by the wayside.

To sum it all up, here is a bit of advice for engineers (or students) and engineering supervisors. It is a severe overstatement to say that this is all there is to it, but this short list touches a lot of issues that arise in engineering day-to-day work.

Engineers:

- However menial or trivial your early assignments may appear, give them your best effort.

- There is always a premium upon the ability to get things done.

- In carrying out a project do not wait for foremen, vendors, and others to deliver the goods; go after them and keep everlastingly after them.

- Confirm your instructions and others' commitments in writing.

- When sent out on any complaint or other assignment, stick with it and see it through to a successful finish.

- Avoid even the appearance of indecision.

- Do not be timid—speak up and promote ideas, but remember other people have good ideas too.

- Before asking for approval of any major action, have a definite plan and program worked out to support it.

- Strive for concision and clarity in oral or written reports.

- Be very careful of the accuracy of your statements.

Engineering Supervisors:

- Every executive must know what is going on in his sphere of authority.

- Do not try to do it all yourself.

- Put first things first, in applying yourself to your job. Establish your priorities, both personal and professional, and keep them.

- Cultivate the habit of boiling matters down to their simplest terms.

- Do not get excited in engineering emergencies; keep your feet on the ground.

- Engineering meetings should not be too large or too small. Get the right people there, but don't waste their time if they don't need to be.

- Cultivate the habit of making brisk, clean-cut decisions.

- Do not overlook the value of suitable preparation before announcing a major decision or policy.

- Plan the work, then work the plan.

- Be careful to freeze a new design when the development has progressed far enough.

- Constantly review developments and other activities to make certain that actual benefits are commensurate with costs in money, time, and manpower.

- Make it a rule to require and submit regular periodic progress reports, as well as final reports on completed projects.

- Do not have too many people reporting directly to one person.

- Assign definite responsibilities.

- Do not create bottlenecks.

So, there it is. Being now informed with this practical guide, engineering is no longer a distant question. You are enlightened in the engineering realm and have inside information of the trade. Hopefully, this will guide you to a greater depth of knowledge within engineering and inspire you to explore the industry further!

THE END

Sample
Patent Drawings

March 12, 1946.   W. A. HITE ET AL   Des. 144,111

AIRPLANE

Filed Oct. 29, 1943   2 Sheets—Sheet 1

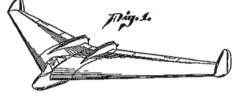

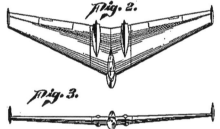

WALTER A. HITE,
RICHARD E. HARVUOT,
DAVID H. WILLIAMS,
INVENTORS.

AIRPLANE PATENT AWARDED TO INTERSTATE AIRCRAFT AND ENGINEERING
CO. FOR AN AIRCRAFT DESIGN DERIVED FROM THAT OF THE XBDR-1

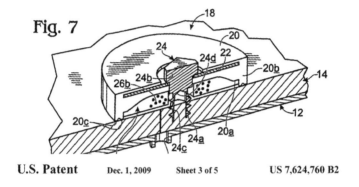

Fig. 7

REPAIR PATENT AWARDED IN 2006 FOR A LIQUID-LEAKING
PUNCTURE WOUND IN THE WALL OF A LIQUID CONTAINER

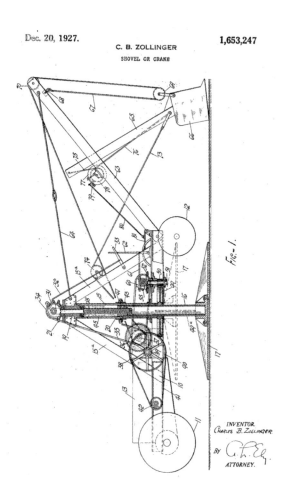

Dec. 20, 1927.

C. B. ZOLLINGER

SHOVEL OR CRANE

1,653,247

PATENT AWARDED IN 1927 TO CHARLES B. ZOLLINGER OF
STERLING, OH FOR A TRACTOR-DRIVEN ROTATING SHOVEL

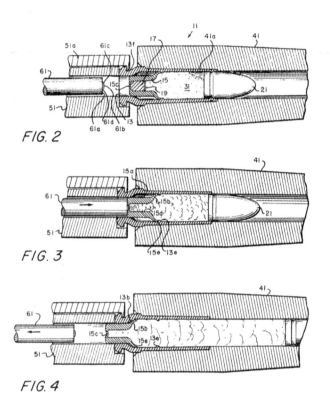

FIG. 2

FIG. 3

FIG. 4

Irwin R. Barr
INVENTOR

BY Reginald F. Pippin, Jr.

ATTORNEY

Piston Primer patent with improved one-piece primer
awarded in 1971 to Irwin R. Barr

Aug. 9, 1949.

S. J. LOYACANO ET AL

Des. 154,788

HOT DOG VENDING CART

Filed Sept. 8, 1948

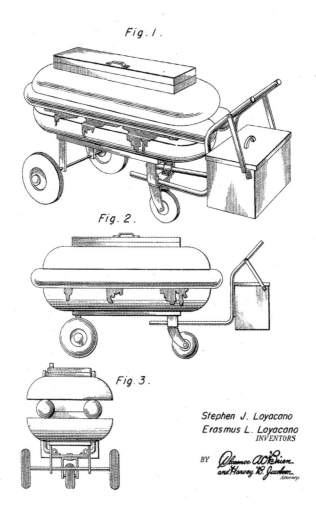

Fig. 1.

Fig. 2.

Fig. 3.

Stephen J. Loyacano
Erasmus L. Loyacano
INVENTORS

BY Clarence A. O'Brien
and Harvey B. Jacobson
Attorneys

PATENT DESIGN OF A HOT DOG VENDING CART AWARDED
IN 1949 TO STEPHEN J. LOYACANO, ET AL.

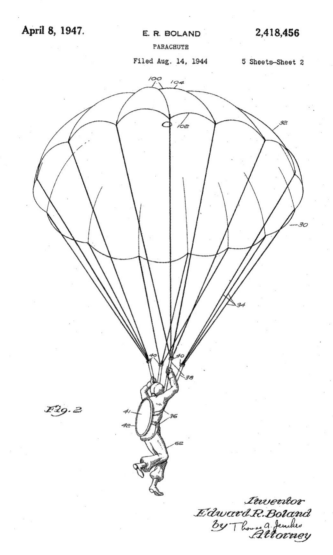

April 8, 1947.   E. R. BOLAND   2,418,456

PARACHUTE

Filed Aug. 14, 1944   5 Sheets—Sheet 2

PARACHUTE PATENT AWARDED TO EDWARD T. BOLAND OF FAIRVIEW, MA IN 1947

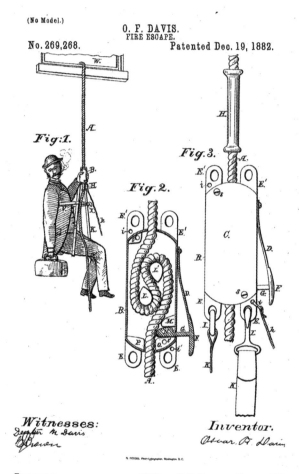

FIRE ESCAPE PATENT AWARDED IN 1882 TO O. F. DAVIS OF TOPEKA, KS.
THE DESCENDING MAN'S INDIFFERENT EXPRESSION WHILE SMOKING A
CIGAR SUGGESTS THAT ESCAPING A FIRE IS A HO-HUM EVENT.

C. M. COOLIDGE.
Processes of Taking Photographic Pictures.

No. 149,724.   Patented April 14, 1874.

WITNESSES.   INVENTOR
Henry N. Miller   Cassius M. Coolidge.
C. L. Ensch.   Alexander Mason
                    By

PATENT AWARDED IN 1874 REGARDING "A PROCESS FOR TAKING A
PHOTOGRAPH OR OTHER PICTURE OF A PERSON'S HEAD LARGE ON A MINIATURE
BODY" TO GIVE AN "EFFECT EITHER COMICAL OR OTHERWISE"

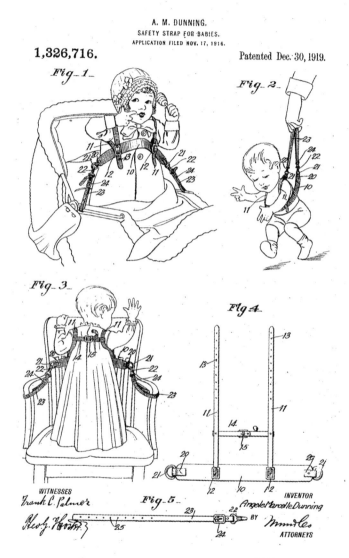

SAFETY STRAP FOR BABIES PATENT AWARDED IN 1919
TO A.M. DUNNING OF SAN FRANCISCO

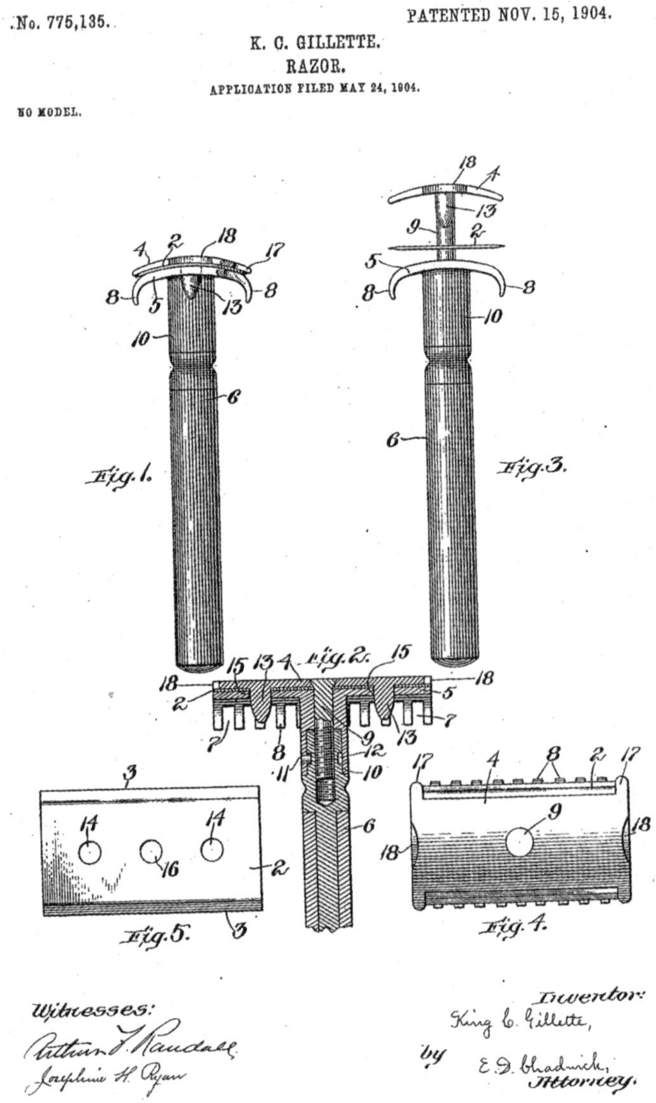

No. 775,135.

PATENTED NOV. 15, 1904.

K. C. GILLETTE.
RAZOR.
APPLICATION FILED MAY 24, 1904.

NO MODEL.

Fig.1.
Fig.2.
Fig.3.
Fig.4.
Fig.5.

Witnesses:
Arthur T. Randall
Josephine H. Ryan

Inventor:
King C. Gillette,
by E. D. Chadwick,
Attorney.

1904 GILLETTE RAZOR PATENT AWARDED TO KING CAMP GILLETTE OF LONDON

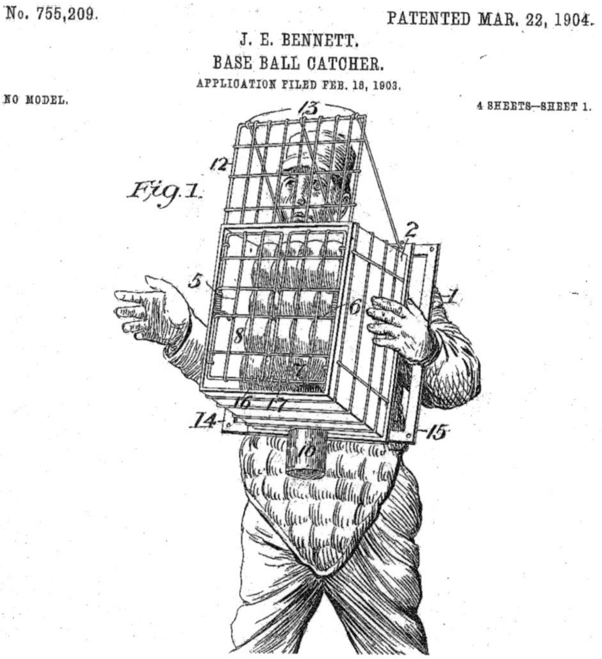

No. 755,209.

PATENTED MAR. 22, 1904.

J. E. BENNETT.
BASE BALL CATCHER.
APPLICATION FILED FEB. 18, 1903.

NO MODEL.

4 SHEETS—SHEET 1.

Fig. 1.

BASEBALL CATCHER PATENT AWARDED TO JAMES E. BENNETT OF MOMENCE, IL IN 1904, TO "RETAIN THE BALL WITHOUT THE PLAYERS' HANDS COMING IN CONTACT THEREWITH"

U.S. Patent   Dec. 18, 1979   Sheet 2 of 2   Des. 253,711

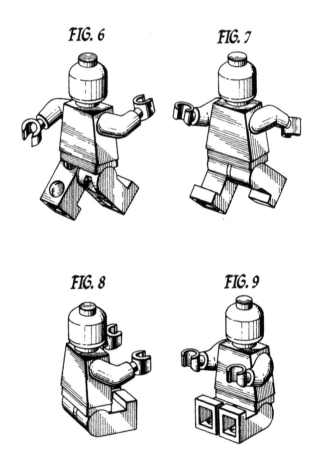

FIG. 6

FIG. 7

FIG. 8

FIG. 9

TOY FIGURE DESIGN BY GODTFRED CHRISTIANSEN, ET AL., OF BILLUND, DENMARK; ASSIGNED TO THE COMPANY INTERLEGO AG OF SWITZERLAND

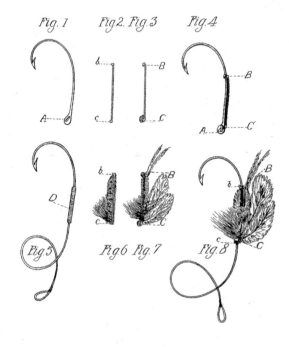

(No Model.)

G. W. UPTON.
FISHING FLY.

No. 468,376.   Patented Feb. 9, 1892.

Fig. 1   Fig 2. Fig 3   Fig 4

Fig 5   Fig 6 Fig 7   Fig 8

Witnesses
E. C. Taylor
E. W. Dow.

Inventor
George W. Upton

FISHING FLY PATENT AWARDED TO GEORGE W. UPTON OF WARREN, OH IN 1892

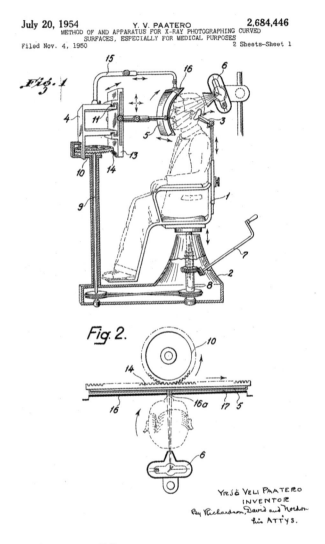

July 20, 1954   Y. V. PAATERO   2,684,446
METHOD OF AND APPARATUS FOR X-RAY PHOTOGRAPHING CURVED
SURFACES, ESPECIALLY FOR MEDICAL PURPOSES
Filed Nov. 4, 1950   2 Sheets—Sheet 1

APPARATUS FOR X-RAY PHOTOGRAPHING CURVED SURFACES PATENT
AWARDED TO YRJÖ VELI PAATERO OF HELSINKI, FINLAND IN 1954

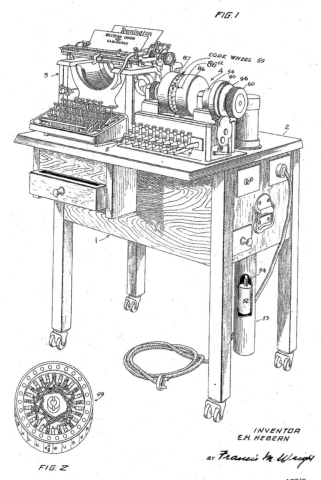

Sept. 30, 1924.

E. H. HEBERN

1,510,441

ELECTRIC CODING MACHINE

Filed March 31. 1921   11 Sheets-Sheet 1

FIG. I

INVENTOR
E.H. HEBERN

BY *Francis M. Wright*

ATT'Y.

FIG. 2

ELECTRIC CODING MACHINE PATENT AWARDED TO EDWARD
H. HEBERN OF OAKLAND, CA IN 1924

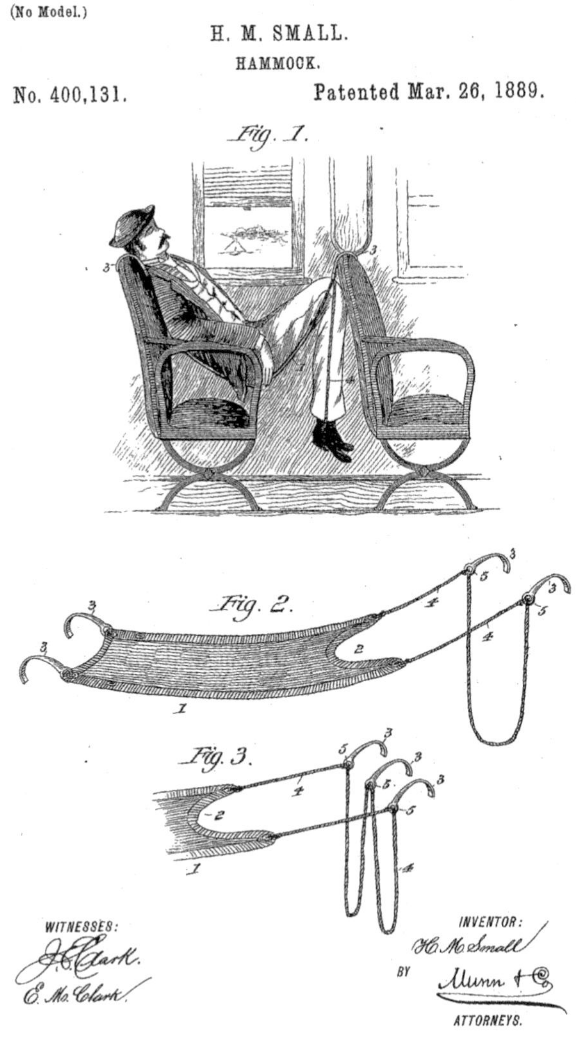

(No Model.)

H. M. SMALL.

HAMMOCK.

No. 400,131. Patented Mar. 26, 1889.

Fig. 1.

Fig. 2.

Fig. 3.

WITNESSES:

INVENTOR:
H. M. Small

BY Munn & Co.

ATTORNEYS.

HAMMOCK PATENT AWARDED TO HERBERT MORLEY SMALL OF BALDWINSVILLE, MA IN 1889, WHEREBY PASSENGERS OF RAILWAY CARS MAY "SLEEP WITH EASE AND COMFORT"

No. 854,364.
PATENTED MAY 21, 1907.

A. O. LOMBARD.
LOG HAULER.
APPLICATION FILED NOV. 22. 1905.

4 SHEETS—SHEET 1.

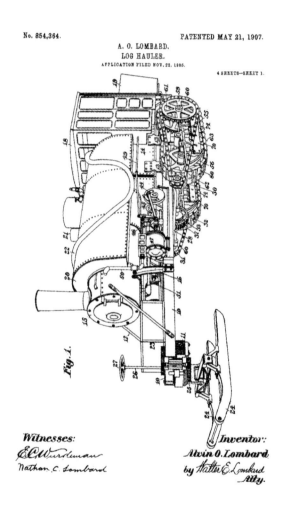

Fig. 1.

Witnesses:

Inventor:

Alvin O. Lombard
by Walter E. Lombard
Atty.

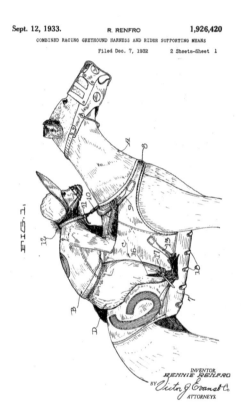

Engineering Quotes

"At that time (1909) the chief engineer was almost always the chief test pilot as well. That had the fortunate result of eliminating poor engineering early in aviation."

IGOR SIKORSKY (RUSSIAN-AMERICAN AVIATION PIONEER)

"Engineers like to solve problems. If there are no problems handily available, they will create their own problems."

SCOTT ADAMS (CREATOR OF THE DILBERT COMIC STRIP)

"We cannot solve our problems with the same thinking we used when we created them."

ALBERT EINSTEIN

"Architects know nothing about everything. Engineers know everything about nothing."

UNKNOWN

"The engineer is a fool who can do for one buck what any other fool can do for two."

UNKNOWN

"Men of initiative and vision who have the ability to draw a good working layout, supervise the drawing of details, check the finished drawings, and then follow the parts in the shop through the assembly and testing phases are the type of all-around engineers constantly being sought to fill positions of higher responsibility."

CHARLES A. CHAYNE (PAST VICE PRESIDENT, GENERAL MOTORS)

"Drafting is probably the engineer's most important form of communication. Just as picture writing, the first form of written communication, was used to express thoughts and records, so do engineers working with graphics mold intangible ideas and theories into tangible goods and products ... there must be design before there can be a test or experimental program ... drafting has a vital, dual function in the engineering organization:

As a means of communication whereby ideas can be presented to others quickly and vividly.

As a tool whereby these ideas may be developed and expanded by graphical methods."

IBID

"'The fewer moving parts, the better.' Exactly. No truer words were ever spoken in the context of engineering."

CHRISTIAN CANTRELL (AUTHOR)

"Engineers ... are not superhuman. They make mistakes in their assumptions, in their calculations, in their conclusions. That they make mistakes is forgivable; that they catch them is imperative. Thus it is the essence of modern engineering not only to be able to check one's own work but also to have one's work checked and to be able to check the work of others."

HENRY PETROSKI (AMERICAN
ENGINEER AND AUTHOR)

"I have not failed, but found 1000 ways to not make a light bulb."

THOMAS EDISON

"A scientist can discover a new star but he cannot make one. He would have to ask an engineer to do it for him."

GORDON LINDSAY GLEGG (ENGLISH
ENGINEER AND AUTHOR)

"The engineer's first problem in any design situation is to discover what the problem really is."

UNKNOWN

"I guess the question I'm asked the most often is: 'When you were sitting in that capsule listening to the countdown, how did you feel?' Well, the answer to that one is easy. I felt exactly how you would feel if you were getting ready to launch and knew you were sitting on top of two million parts—all built by the lowest bidder on a government contract."

JOHN GLENN (US SENATOR,
AVIATOR, ASTRONAUT, ENGINEER)

"If it's not broken, take it apart and fix it anyway."

UNKNOWN

"Architects and engineers are among the most fortunate of men since they build their own monuments with public consent, public approval and often public money."

JOHN PREBBLE (ENGLISH JOURNALIST)

"The trick to having good ideas is not to sit around in glorious isolation and try to think big thoughts. The trick is to get more parts on the table."

STEVEN JOHNSON (AMERICAN AUTHOR)

"I've never seen a job being done by a five-hundred-person engineering team that couldn't be done better by fifty people."

C. GORDON BELL (AMERICAN ENGINEER)

"One has to watch out for engineers—they begin with the sewing machine and end up with the atomic bomb."

MARCEL PAGNOL (FRENCH NOVELIST), *CRITIQUES DES CRITIQUES*

"When engineers and surveyors discuss aesthetics and architects study what cranes do, we are on the right road."

OVE ARUP (DANISH-ENGLISH ENGINEER)

"The ideal engineer is a composite ... he is not a scientist, he is not a mathematician, he is not a sociologist or a writer, but he may use the knowledge and techniques of any or all of these disciplines in solving engineering problems."

N.W. DOUGHERTY (1940-56 DEAN OF ENGINEERING, U. OF TENNESSEE)

"Engineering stimulates the mind. Kids get bored easily; they have got to get out and get their hands dirty: make things, dismantle things, fix things. When the schools can offer that, you'll have an engineer for life."

BRUCE DICKINSON (ENGLISH MUSICIAN AND PILOT)

"Thirty spokes share the wheel's hub; it is the center hole that makes it useful. Shape clay into a vessel; it is the space within that makes it useful. Cut doors and windows for a room; it is the holes which make it useful. Therefore, benefit comes from what is there; usefulness comes from what is not."

EXCERPT FROM *TAO TE CHING*

"The goal of science and engineering is to build better mousetraps. The goal of nature is to build better mice."

UNKNOWN

"I would rather teach engineers than anything. They're so bad to start with, it's a great joy to see them learn how to open their mouths without chewing their tongues up."

DALE CARNEGIE

The
Practical
Alternative

Who Says

Engineering is Boring?

I N 1984 NASA D<small>RYDEN</small> F<small>LIGHT</small> R<small>ESEARCH</small> F<small>ACILITY</small> and the Federal Aviation Administration (FAA) teamed up in a unique flight experiment called the "Controlled Impact Demonstration," or CID. The test involved crashing a Boeing 720 aircraft with four JT3C-7 engines burning a mixture of standard fuel with an additive designed to suppress fire. In a typical aircraft crash, fuel spilled from ruptured fuel tanks forms a fine mist that can be ignited by a number of sources at the crash site. The anti-misting kerosene (AMK) additive, when blended with Jet-A fuel, had demonstrated the capability to inhibit ignition and flame propagation of the released fuel in simulated crash tests. In addition to the AMK research, the NASA Langley Research Center was involved in a structural load measurement experiment, which included having instrumented crash-test dummies filling the seats in the passenger compartment.

The final flight on December 1, 1984 was preceded

by more than four years of effort trying to set up final impact conditions considered survivable by the FAA. During those years, while fourteen flights with crews were flown, the following major efforts were underway: NASA Dryden developed the remote piloting techniques necessary for the B-720 to fly as a drone aircraft; General Electric installed and tested four degraders (one on each engine); and the FAA refined AMK (blending, testing, and fueling a full-size aircraft). These flights were used to introduce AMK one step at a time into some of the fuel tanks and engines while monitoring the performance of the engines.

On the fifteenth flight, which had no crew, all fuel tanks were filled with a total of 76,000 pounds of AMK. The remotely piloted aircraft landed on Rogers Dry Lakebed in an area prepared with posts to test the effectiveness of the AMK in a controlled impact. The CID was spectacular, with a large fireball enveloping and burning the B-720 aircraft. From the standpoint of AMK the test was a major setback, but for NASA Langley, the data collected on crashworthiness was deemed successful and just as important.

PHOTO CREDITS

Cover Photo: *Flying Automobile,* Jess Dixon, November 1941. (Modern Mechanix) Source: Wikimedia Commons

p. x: *7630 Mountain and Mining Transit,* 1917. (Kolesch & Company) Source: Wikimedia Commons

p. 1: *Ball Compound Steam Engine,* New Catechism of the Steam Engine, 1904. Source: Wikimedia Commons

p. 9: *Engineering students and professor from USC surveying 4th St. in downtown Los Angeles,* 1912. The professor can be seen in the right foreground, wearing a derby hat, standing next to a student looking through what appears to be a theodolite mounted on a tripod. The student can be seen looking through an eyepiece on the theodolite towards a second student at far left, who can be seen holding a long measuring stick in front of the Fremont Hotel, which can be seen in the background. A third student can be seen standing, center, holding a large piece of paper. Source: Wikimedia Commons

p. 15: *Mechanical Drawing Class,* Tuskegee Institute, 1902. (Library of Congress)

p. 17: *Perpetual Motion,* Norman Rockwell, 1920. (Popular Science) Source: Wikimedia Commons

p. 19: *Engineers Demonstrating the Cantilever Bridge System,* 1887. *(Rare Historical Photos)*

p. 22: *Caveman Cartoon,* Matthew Diffee, Cartoon Collections CC123941, www.cartooncollections.com.

p. 25: Henry Ford with Model T, Hotel Iroquois, Buffalo, NY, 1921. Source: Wikimedia Commons

p. 26: *We Can Do It,* J. Howard Miller, 1943. (U.S. National

Archives and Records Administration) Source: Wikimedia Commons

p. 29, 31-32: *Professional Engineering Stamps*, www.pestamps. com. Used by Permission

p. 39: *Herbert Hoover*, George Grantham Bain Collection. (Library of Congress) Source: Wikimedia Commons

p. 44: *Charles Lindbergh at Ryan Aircraft*, San Diego, May 1927. (Missouri Historical Society)

p. 45: *Motorcycle*, William S. Harley et al., Patent No. 1,510,937, 7 October 1924. Source: Wikimedia Commons

p. 50: *New Iron Horse Tricycle*, Charles W. Oldreive, 1882. Source: Wikimedia Commons

p. 51: *Warren Safety Helmet*, 1912. (Flight magazine)

p. 52: *Hot Air Balloon*, E. Watson, 2013. Source: Wikimedia Commons

p. 59: *Dilbert Minimaliste*, E. Ripounet, 2006. Source: Wikimedia Commons

p. 67: *Bridges 28*, 1911. (Encyclopedia Britannica) Source: Wikimedia Commons

p. 74: *Transmission of Motion by Compound Gear Train*, 1911. (Army Service Corps Training, Mechanical Transport) Source: Wikimedia Commons

p. 75: *Boeing Vertol 107-III Certification Team*, Charles Kessler. Charles Kessler is a retired flight test engineer for Boeing's Vertol helicopter division (formerly Piasecki Helicopter Co.). He joined Piasecki in 1947, in the company's fourth year, and retired from Boeing in 1983. During his 37-year career he took part in the testing of prototypes and alterations of such models as the CH-47 Chinook and Sea Knight, the H-16, HRP-2, and the V-107. He taught the stability augmentation system to the German Luftwaffe. He has written about his experience in the blog "Early Helicopter Years," which can be found at helicopterstory. blogspot.com.

p. 79: *Cityscape Image: www.freepik.com.*

p. 79: *Simple Flourish* #2418940, Clip Art Library, www.clipart-library.com.

p. 106-107: CID *Aircraft Test* (Series), NASA Glenn Research Center, 1984.

The following artists provided icons from flaticon.com website: Pixel perfect, flaticon, Freepik, Grego Cresnar, Icongeek26, xnimrodx, Eucalyp, itim2101, Nikita Golubev, Good Ware, Skyclick, turktub, Pause08, DinsoftLabs, sript, Smashicons, dmitri13, Alfredo Hernandez, surang, Darius Dan, Becris, inconixar, and smallikeart.

REFERENCES

p. 1: "engineer", "ingenerare", "ingenium", Ralph J. Smith, Encyclopedia Britannica, 2017.

p. 27: "Good Engineer / Bad Engineer", Kevin Beauregard, Ben Horowitz https://hackernoon.com/good-engineer-bad-engineer-c6e2dea98b9b

p. 29: NCEES; https://ncees.org/

p. 30: ABET; www.abet.org

p. 30-34: NSPE; www.nspe.org

p. 33-35: *Engineer Your Way to Success*, Shawn P. McCarthy, NSPE, 2000.

p. 36-37: *Mechanical Engineering Reference Manual*, 13th ed., M. Lindeberg, 2013.

p. 37:; "Are Engineering Fees a Percentage of Construction Costs?", Jayne Thompson, July 2020, https://smallbusiness.chron.com

p. 39-40: Herbert Hoover Presidential Library Association

p. 45: "a legal right...", Cambridge Dictionary, Cambridge University Press, dictionary.cambridge.org/us/dictionary/English/patent

p. 46-49: United States Patent Office; www.uspto.gov/patent

p. 52-53: "A man is flying in a hot air balloon...", www.et.byu.edu/~tom/jokes/ENGINEERS_&_MANAGERS.html

p. 54: *Why Buildings Stand Up*, Mario Salvadori, 1980.

p. 55: "Dilbert's Salary Theorem", Charles Jeon, University of Pennsylvania. Used by Permission.

p. 59: "The Engineer Personality", Dr. Charles E. Goshen, 1954.

p. 61: "I Married an Engineer," B. Smith, 1973. (Marriage Magazine)

p.65: "Normal people...", www.cmrr.umn.edu/~strupp/engr.html

p. 76-78: "The Unwritten Laws of Engineering", W.J. King, 1944. (Mechanical Engineering)

p. 97-101: Selected Quotations, https://www.workflowmax.com/blog/101-engineering-quotes-from-the-minds-of-innovators

p. 105-107: Controlled Impact Demonstration, NASA Glenn Research Ctr. www.nasa.gov

LUKE ZOLLINGER IS A LICENSED PROFESSIONAL ENGIneer in mechanical and civil engineering (OR #87563PE). His engineering experience includes tandem rotor helicopters, concrete, steel, and wood bridges, heavy construction, structural engineering, and airport design. He currently is the chief engineer for a research and development company, which involves ballistic tests, blast testing, and other unconventional things. Mr. Zollinger was born at a very young age, subsequently graduated from engineering school, and now lives with his wife and five children.